APUN
The Arctic Snow

Matthew Sturm

University of Alaska Press
Fairbanks, Alaska

Dedicated to

Arnold Brower, Sr.

of Barrow, Alaska, who understood snow.

Arnold Brower, Sr., and the author in 2008. Photo by Craig George.

University of Alaska Press
P.O. Box 756240
Fairbanks, AK 99775-6240

ISBN 978-1-60223-069-9

Printed in Canada

Library of Congress Cataloging-in-Publication Data

Sturm, Matthew.
Apun, the arctic snow / by Matthew Sturm.
p. cm.
ISBN 978-1-60223-069-9 (pbk. : alk. paper)
1. Snow—Arctic regions—Juvenile literature. I. Title.
QC926.37.S78 2009
551.57'842113—dc22

2009016373

This publication was printed on acid-free paper that meets the minimum requirements for ANSI / NISO Z39.48–1992 (R2002) (Permanence of Paper for Printed Library Materials).

Text and cover design by Paula Elmes, ImageCraft Publications & Design

Cover illustrations by Ken Libbrecht

Contents

Snow
and
Life

It's October.

It's getting colder.

Clouds are building over

the ocean north of

Barrow, Alaska.

Barrow: the farthest north town in the U.S.

2

Snow and Life

Soon it will begin to snow. Temperatures will drop and the long winter night will start. Through the night, layer by layer, the snow cover will grow deeper. The snow crystals in each layer will change as the winter passes. The sun will come back in January, and by May the snow will start to melt. The brief summer will begin, but the snow won't be gone for long because next October the cycle will start again. This is the life-cycle of the arctic snow cover—*apun* in Iñupiaq.

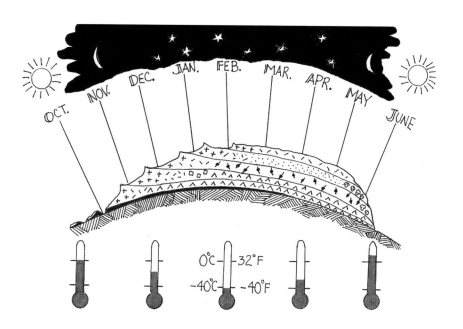

A *pun* can mean life or death for the plants and animals of the Arctic. For example, the snow cover is the blanket that keeps the lemmings warm in the winter, a soft and fluffy down quilt that they can snuggle under!

The lemmings also use the snow cover for protection. They build tunnels in the snow to live in and to store their food. Most of the winter, they are under the snow.

But if a white weasel, which eats lemmings, gets in the tunnels, well, then it is a terrifying place where death can lurk around the next corner.

S ome creatures like the caribou and wolf live above the snow, but it is still important to them.

hoof-print in
the snow

The caribou have long legs for running through deep snow and split hooves that spread out and support their weight.

Wolves and foxes have big paws so they can run on top of the snow and chase prey.

paw-print in
the snow

Snow
and
Life

E ven the plants need the snow! The snow keeps their roots warm and protects their branches from drying out in the harsh wind. It protects the branches from being "sanded" to death by blowing snow, which in Iñupiaq is *natiġvik*. Just like the lemmings, some plants like being under the snow.

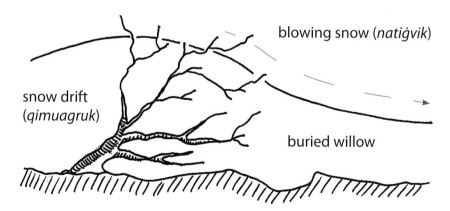

blowing snow (*natiġvik*)

snow drift (*qimuagruk*)

buried willow

Apun, the arctic snow cover, means good traveling for people—and fun too!

8

Snow Crystals

A pun starts in a cloud! But even though it is winter and the weather has turned cold, the water in the clouds does not want to freeze into snow crystals. It needs help.

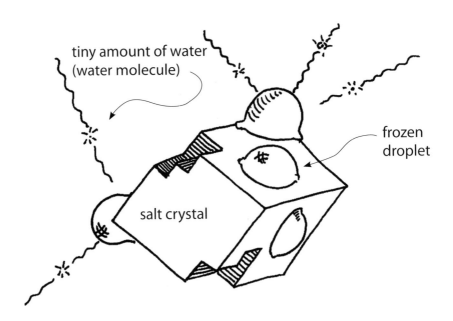

tiny amount of water (water molecule)

frozen droplet

salt crystal

Fortunately, particles of dust and salt from the sea, which are also in the cloud, help the water freeze. The particles are tinier than the point of pin, but the water molecules are even smaller.

Snow Crystals

As the cloud droplets freeze around dust and salt particles, tiny ice crystals form. Five crystals could fit on the point of a pencil. They are so small and light, they float in the air instead of falling to the ground. They will need to grow a lot before they can pile up on the ground as apun. As they float, they sparkle in the sun like diamonds.

This is called diamond dust, or in Iñupiaq, *irriqutit*.

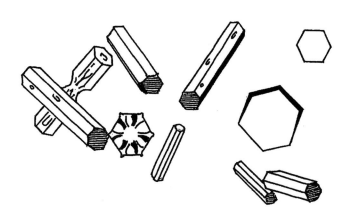

The clouds over the ocean are moist. The diamond dust is blown up and down in the moist clouds. The crystals eventually grow into snowflakes (*qannik*) heavy enough to fall out the bottom of the cloud. The falling crystals have perfect hexagonal shapes and funny names.

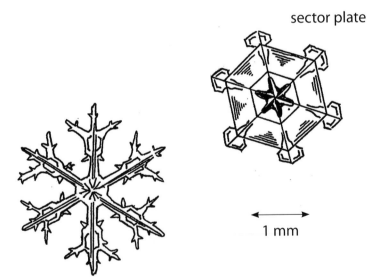

sector plate

1 mm

stellar dendrite

Snow Crystals

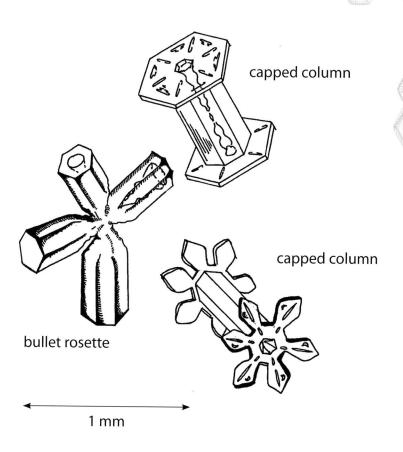

capped column

capped column

bullet rosette

1 mm

Some snowflakes are hexagonal but don't look anything like stellar dendrites or sector plates.

The perfect shapes of these crystals happens because they are made of water molecules which look like this:

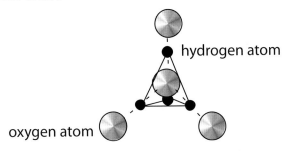

hydrogen atom

oxygen atom

When several water molecules are linked together, it looks like this:

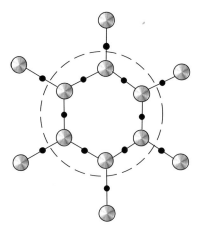

Can you see why a snowflake would have six arms and a hexagonal shape?

Changing
Snow

The first thing that happens in October is that fluffy new snow (*nutaġaq*) falls from the sky and covers the tundra. The tundra is likely to still be warm from the summer, so this first snow often melts immediately (*auktuq*), leaving the tundra a patchwork of bare tundra and wet snow.

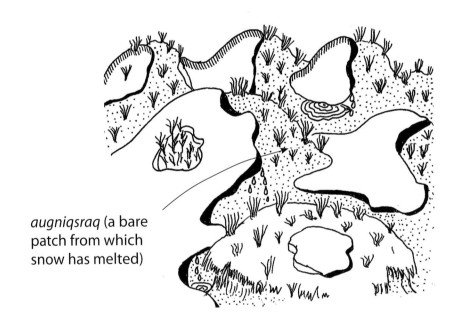

augniqsraq (a bare patch from which snow has melted)

Changing Snow

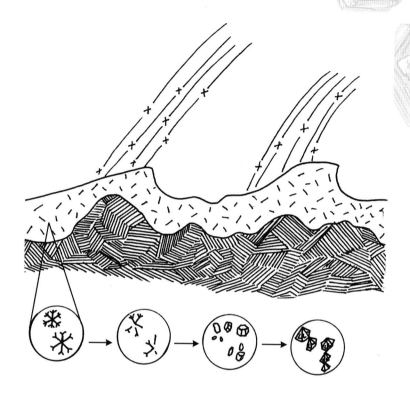

S oon, however, it gets colder and the snow no longer melts. The tundra gets completely covered and looks white. But unseen, down in the snow and at the snow surface, the snow crystals continue to change. This process has a fancy name: "snow metamorphism." The snowflakes change into something quite different by the end of the winter.

There are three forces that alter and change snow crystals:

Wind (*anuġi*)

Temperature Gradient
(*qanuġinniŋa siḷam*)

Heat (*uunnaq*)

Wind

Soon, however, the tundra cools off, and the arctic wind (*anuġi*) starts to blow. It is the time of the blizzards. The wind blows and tumbles the newly fallen snowflakes, breaking them into pieces. If you were to watch closely when the wind is driving the snow, it would look like thousands of tiny white grasshoppers or *miññuq* (small arctic beetles) jumping and bounding along, crashing into each other, and jumping again.

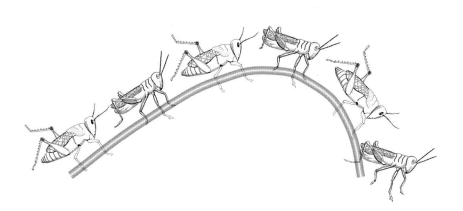

Of course, it isn't really white grasshoppers or beetles: it's jumping grains of snow, and there is a trick to how the wind moves these grains. It is called saltation.

One snow grain will be driven by the wind into second grain on the ground.

This will bounce the second grain up into the air!

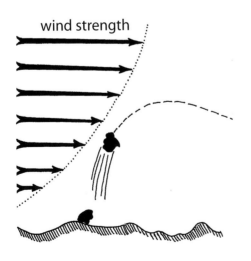

wind strength

The wind is stronger higher up, so the second grain goes flying down the wind, only to crash into a third grain, and so on.

Changing Snow

There are many words in Iñupiaq for blowing snow.

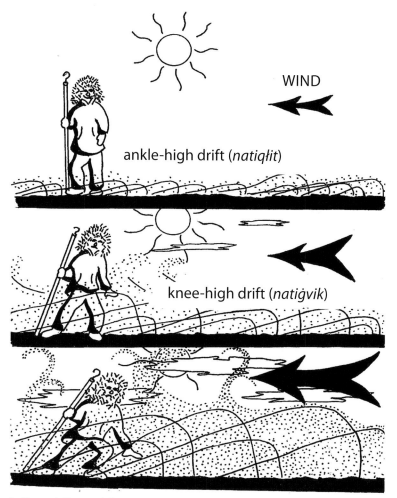

WIND

ankle-high drift (*natiqłit*)

knee-high drift (*natiġvik*)

full-on blizzard (*agniqsuq*)

When the wind stops blowing, the broken snow grains become "glued" together in a super-strong mass. The glue joints are called bonds, and the thicker they are, the stronger the snow. This process is called sintering. Some snow is so hard it can only be cut with a saw. Depending on how hard the snow is, it might be called *aniu* (softer), *aniuvak* (harder), *nuturuk* (firm, good for making a snow house), or *siḷḷiqsruq* (super-hard, often icy).

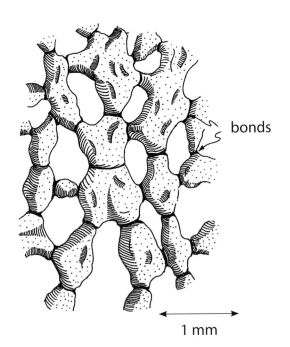

bonds

1 mm

Changing
Snow

The wind creates fantastic drifts and snow shapes.

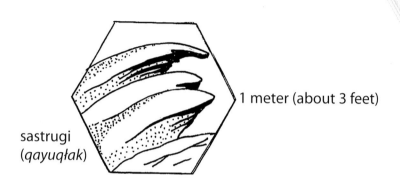

sastrugi
(*qayuqłak*)

1 meter (about 3 feet)

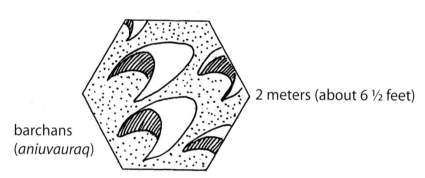

barchans
(*aniuvauraq*)

2 meters (about 6 ½ feet)

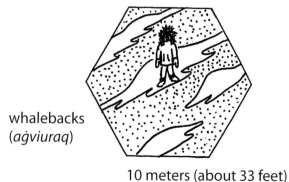

whalebacks
(*aġviuraq*)

10 meters (about 33 feet)

The drifting snow fills in gullies and cut-banks. More snow is added during each new blizzard.

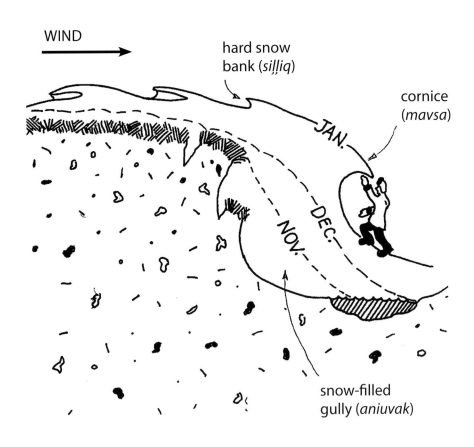

WIND →

hard snow bank (*siḷḷiq*)

cornice (*mavsa*)

JAN.

DEC.

NOV.

snow-filled gully (*aniuvak*)

Temperature Gradient

What is a temperature gradient? Imagine it is cozy and warm inside (fire), but frigid outside (icicles). One side of the wall is warm, the other cold. The temperature gradient is the difference between the warm and cold air. It moves moisture from inside to outside of the house.

+20° -40°C

20 cm

What does a temperature gradient do to snow?

It makes water molecules move from warm snow grains (which are nearer the ground) to cold ones (which are nearer the air).

The bottoms of the grains grow fast. The tops of the grains disappear (evaporate).

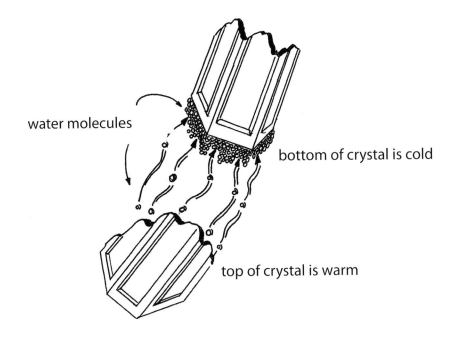

water molecules

bottom of crystal is cold

top of crystal is warm

Changing Snow

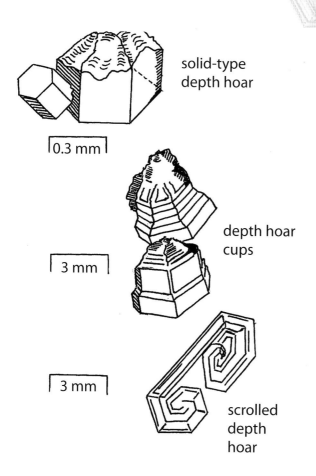

solid-type
depth hoar

| 0.3 mm |

depth hoar
cups

| 3 mm |

| 3 mm |

scrolled
depth
hoar

These fast-growing crystals develop beautiful geometric shapes. They are called depth hoar or *pukak*.

The first snow of the year (*apivaalluqqaaġniq*) filled in the holes between tundra tussocks. It has been there since October. As more snow has fallen on top of it, the bottom layers have gotten warmer. Driven by a temperature gradient, the snow grains have now grown into depth hoar crystals. This layer is also called *pukak* in Iñupiaq.

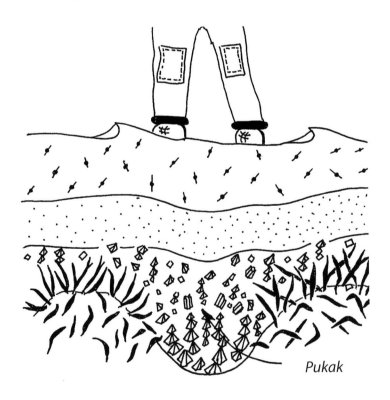

Pukak

Changing Snow

Wouldn't it be fun if the *pukak* (depth hoar) crystals were big and we could explore them?

While we may not be able to crawl through and climb on the crystals, if you place a few of them in your hand and look at them with a magnifying glass, it is almost as good.

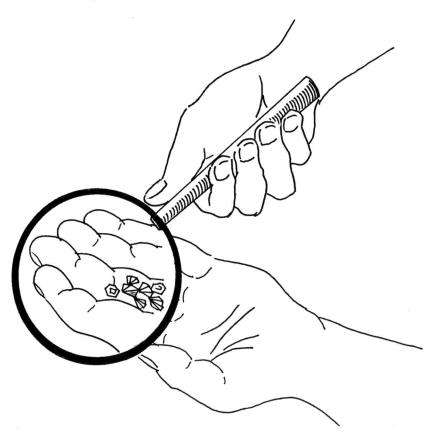

Heat

Changing Snow

Finally a day comes when the sun comes out bright and strong. The air is warm and still. There is a strange sound, one that hasn't been heard since last spring: drip, drip, drip.

The heat makes the snow melt, which produces snow grains with rounded shapes. The necks (bonds) between grains thicken but for different reasons than wind sintering. Melt water collects where the ice balls touch.

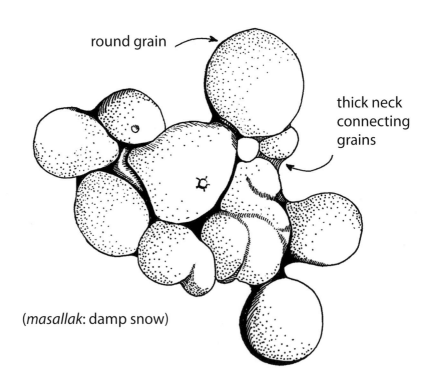

round grain

thick neck connecting grains

(*masallak*: damp snow)

Changing Snow

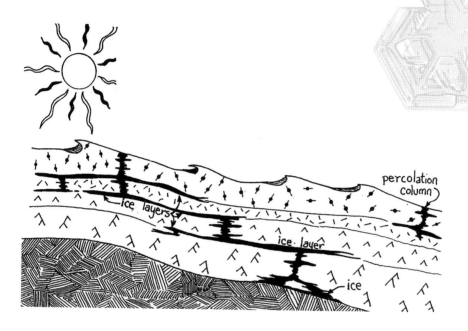

If there is a lot of melting, then melt water will "percolate" down into the snow and spread out along snow layers, where it will refreeze and make ice layers. A surprising trick is that the melt water can make several ice layers at one time, the layers connected by knobby pipes called percolation columns. Can you find these?

Sometimes just a little melting will take place right at the surface. Perhaps it will be warm enough for mist or even rain. The snow surface will become hard and shiny. In Iñupiaq this is called:

siḷḷiġruaq

The Germans have a nice word for this:

firnspiegel

It means ice mirror.

Slush

immaktinniq

When snow is melting, the last stage of melt, before it is all water, is slush. A slush is a mixture of water and ice, like ice cubes in a glass of water, except the cubes are slightly flattened balls only 5 to 10 millimeters across, about the width of a pencil eraser.

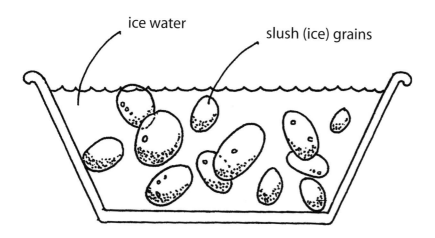

ice water   slush (ice) grains

35

Snow
and
People

Snow and People

The arctic snow cover has been important to northern people for centuries. Before the Iñupiat had metal, they made snow knives (*panak* or *saviuraqtuun*) from bone and ivory. Some bone knives are more than 600 years old! Today, snow knives continue to be used, but they are made from steel. They are still used to cut and shape snow, because even today, working with *apun* means survival in the North.

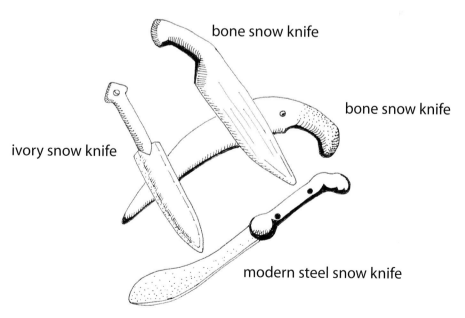

bone snow knife

bone snow knife

ivory snow knife

modern steel snow knife

One use of snow knives is to make shelter. Good snow houses (*aniquyyaq*) must be built from strong, deep snow (*aniuvak*). The snow blocks must be cut with skill so that the shelter won't blow down.

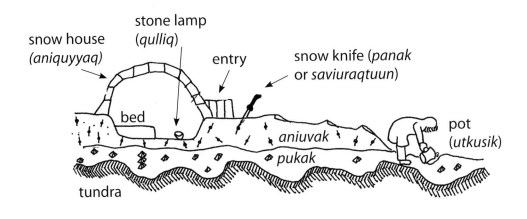

snow house (*aniquyyaq*) — stone lamp (*qulliq*) — entry — snow knife (*panak* or *saviuraqtuun*) — bed — aniuvak — pukak — pot (*utkusik*) — tundra

The builder must take care not to dig into the *pukak* at the bottom of the snow or else the wind will blow through the snow into the house. The loose *pukak*, however, is good for filling a pot to melt snow for water because it is weak and scoops up easily. The house blocks are glued together with loose snow that sinters (remember?) one block to another.

Snow and People

Apun is particularly important now that the climate is warming and the sea ice is disappearing. What does *apun* have to do with sea ice? Everything!

The snow on the ice reflects sunlight, slowing the rate of melting. It also insulates the ice from cold winter temperatures, slowing the rate the ice thickens by freezing.

Which will win? Reflecting or insulating? Can *apun* save the ice?

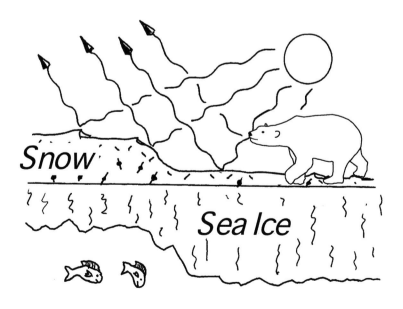

Snow

Sea Ice

For nine months of the year

  Apun

    is Life in the

      Arctic.

Iñupiaq Glossary

Agniqsuq. a full-on dry snow blizzard, during which it is hard to see anything because it is a white-out (*quvyugaġnaq*). See also *piqsiqsuq*: a wet snow blizzard).

Aġviuraq. a whaleback drift (possibly spelled *avoorik*).

Aniquyyaq. a snow house or igloo.

Aniu. packed snow.

Aniuvak. hard packed snow in a gully. A big drift in the lee of a cabin.

Aniuvauraq. a drift with a sharp downwind side and a smoother upwind side; a crescent-shaped drift called a barchan.

Anuġi. wind.

Apivaalluqqaaġniq. first snow of the year in October.

Apun. the snow cover, a general term for snow on the ground.

Augniqsraq. a patch of tundra from which snow has melted.

Auktuq. melting snow.

Immaktinniq. slush; very wet snow, as in an area where river water flowed into the snow (a snow swamp).

Irriqutit. diamond dust or ice crystals in the air (which means a cold spell is coming).

Masallak. snow damp enough to stick, as in making snowballs; also spelled *masayyak*).

Mavsa. an overhanging snow cornice (in Kotzebue: *mapsa*).

Miñŋuq. a little jumping beetle, sometimes red (comes from the word minŋiq, which means to jump up or over).

Natiġvik. low blowing snow, with drift no higher than the knee.

Natiqłit. even lower drifting snow than *natiġvik*.

Nutaġaq. fresh powder snow.

Nuturuk. good material to make a snow house, firm yet not too hard.

Panak. a snow knife (Canadian: see *saviuraqtuun*).

Piqsiqsuq. a wet snow blizzard.

Pukak. coarse, large ornate depth hoar crystals making up a layer at the base of the snow cover.

Qannik. a snowflake.

Qanuġinniŋa siḷam. the drop in temperature across a house wall or the snow pack; a temperature gradient.

Qayuqłak. an anvil-head drift or sastrugi.

Qimuagruk. a long, built-up snowdrift similar to a whaleback (*aġviuraq*), useful for navigation.

Qulliq. a stone lamp.

Saviuraqtuun. a tool (i.e., knife) used by a man, possibly a snow knife (see *panak*).

Siḷḷiġruaq. slippery icy snow conditions, as a surface ice layer.

Siḷḷiq. hard icy snow above *pukak*, sometimes the product of a rain-on-snow event.

Siḷḷiqsruq. old icy snow, extra-hard. So hard no snowmachine or animal tracks are left on it. Hard to dig with a shovel. The snow surface can be a lot like lake ice. Looks like *firnspiegel* (ice mirror).

Utkusik. a cooking pot.

Uunnaq. heat.

Acknowledgments

Many people have helped make this book possible. Larry Kaplan, Arnold Brower Sr., Kenny Toovak, Craig George, Martha Stackhouse, and Fanny Akpik helped with the Iñupiaq glossary. Glenn Sheehan of the Barrow Arctic Science Consortium supported the work in many ways. Carl Benson, Sam Colbeck, Chuck Racine, and Don Perovich taught me much about snow. Jon Holmgren, Walt Tape and Ken Libbrecht were particularly generous with their snow crystal photographs. Sue Mitchell and her production team at the University of Alaska Press made the book look beautiful. Finally, my wife, Betsy, tested the book on her second grade class, and has always encouraged my writing and drawing.